YOUR KNOWLEDGE HAS VALUE

- We will publish your bachelor's and
  master's thesis, essays and papers

- Your own eBook and book -
  sold worldwide in all relevant shops

- Earn money with each sale

Upload your text at www.GRIN.com
and publish for free

GRIN

Wine glasses as multiphonic instruments forced to sound like monophonic instruments

Alexander Rehm

Bibliographic information published by the German National Library:

The German National Library lists this publication in the National Bibliography; detailed bibliographic data are available on the Internet at http://dnb.dnb.de.

ISBN: 9783346838896
This book is also available as an ebook.

© GRIN Publishing GmbH
Nymphenburger Straße 86
80636 München

Print and binding: Books on Demand GmbH, Norderstedt, Germany
Printed on acid-free paper from responsible sources.

The present work has been carefully prepared. Nevertheless, authors and publishers do not incur liability for the correctness of information, notes, links and advice as well as any printing errors.

GRIN web shop: https://www.grin.com/document/1337700

Wine glasses: Multiphonic instruments forced to sound like monophonic instruments!

A. Rehm

Summary

The acoustic parameters of various types of "singing wine glasses" have been investigated after a finger-excitation (sliding the finger on the rim of the wine glass) or a hit-excitation (short hit with a metal stick against the bowl of the wine glass). After a hit-excitation, the wine glasses show typical characteristics of a multiphonic sounds whereas a finger-excitation generates a monophonic sound known from monophonic instruments. It could be confirmed that filling water in the bowl will lower the frequency of the base-tone and the respective harmonics as described previously, but it could also be demonstrated that liquid in the bowl will change the intensity as well as the timbre of the generated sound significantly, which should be considered by musicians using glass harps. Under certain circumstances of a finger-excitation, the resulting sound shows in parallel features of a monophonic sound (base-tone and related harmonics) and features of a multiphonic sound (frequencies with no harmonic relationships plus complex tones). As the "construction" of a wine glass is by far less complex than a wood wind instrument, it might be a good object to investigate the principle mechanism of the production of complex tones within multiphonic sounds generated by "tube-like" instruments as wood winds.

Table of Contents

Introduction

There are various shapes of wine glasses which are classified according to the wine or liquor which are supposed to be drunk – see Table 1 (Ref: 1)

Table 1 also demonstrates that for drinking purposes in most cases the bowls are filled only half or even less. This habit gives the opportunity that the glasses will generate a certain sound if two or more persons execute the typical "cheers-procedure" by shortly hitting their glasses against the others. It is common knowledge that the generated sound depends mainly on the glass-characteristics (e.g. shape, glass-structure, thickness of the walls) and the filling status. Experienced musicians easily recognize that the sound generated by hitting two glasses against each other is a multiphonic sound exhibiting different prominent non-harmonic frequencies. Several investigations and recent studies have analyzed the sound and the movement of wine glasses after an excitation and have been able to describe parameters defining the Eigenfrequenzen of wine glasses (Ref. 2-9).

Wine glasses are also used by musicians as parts of glass harps. In these cases, the sound is generated by moving the finger on the rim of the bowl of the glass. The result is a clear monophonic sound existing of a prominent frequency defining the basic frequency (pitch of the tone) and several harmonics (multiples of the basic frequencies) of lower intensity. In most cases, the basic frequency generated in such a way is very close to the first prominent Eigenfrequenz after an excitation of the glass by a hit.

In this study, the differences in the sounds of wine glasses stimulated either through a hit against the bowl or through a finger movement on the rim of the bowl will be investigated in detail. Further, it will be analyzed how the filling of the bowl with liquid will influence the sound of the wine glass and whether wine glasses may generate similar multiphonic sounds as known from wood wind instruments (Ref. 10, 11).

Anmerkung der Redaktion: Die Abbildung wurde aus urheberrechtlichen Gründen entfernt.

Table 1 – Types of wine glasses (Ref: 1)

Material & Methods

Several commercially available wine glasses have been used for this study. According to the types of glasses as described in table 1 the following wine glasses have been investigated: "white", "Alsace", "Hock", "large Bordeaux", "Pinot noir" and "Cabernet Sauvignon" (see figure 1). All measurements (recordings of sounds) were done under quiet conditions at a stable temperature of 20 degrees Celsius. Even the water filled in the chalice of the wine glasses have been kept at this temperature.

The recording equipment consists of a Behringer Xenyx 1204 mixer and a Rode NT5 microphone. The recording software was Mixcraft 8 and the analysis software was Praat (Ref. 12).

The analysis software Praat allowed to measure and calculate the following data from wav-files with a sampling rate of 44,1kHz:

a) Power spectra (frequency-spectra) of sounds derived from a Fast Fourier Transformation procedure (FFT)
b) Reverse FFT procedure to generate wave-files for certain frequency-ranges in the power spectrum.
c) Long-Term Average Spectra (Ltas). A Ltas is a FFT-generated power spectrum of the frequencies comprising a sound sample. Praat gives the option to define the frequency bandwidth of the analysis, which is important as this defines the resolution of the Ltas.
d) Data on the amplitude (in Pa), the intensity (in dB), the power (in Pa^2) or the energy (in Pa^2sec) of the investigated oscillation.
e) Determination of the "Center of Gravity" in Hz (using a 2.power calculation) of a power spectrum as a parameter to describe and compare the timbre of different sounds.

Figure 1: Example of a wine glass of type "Alscace" or "Cabernet Sauvignon" used in this study.

The excitation procedure of the wine glasses to produce a sound was the following:

a) Hit-excitation

A hit against the bowl of the glass was performed with a metal stick in such a way that the stick received horizontal movement energy, so it could swing against the bowl and bounce back after the hit. For wine glasses of the type "Alsace" and "Cabernet Sauvignon" the hit was performed at the half height of the bowl. In case of other wine glasses it was performed at the point of the largest diameter of the bowl. For each experiment it was ensured that the hit took place at the same point of the bowl as a deviation influences the excitation of the oscillation modes (Ref 5, 8). The metal stick used for the experiments showed no Eigenfrequenz in the range of 0-20.000Hz, which was the measurement range for all experiments.

b) Finger-excitation

Before a measurement was performed with a certain wine glass, the manipulating person evaluated the ideal pressure of the finger on the rim of the bowl as well as the ideal speed of the finger movement on the rim to produce the most stable sound. After this evaluation and some minutes of testing to gain a high repeatability in producing a stable sound, the recordings have been executed.

For both excitation procedures, it is obvious that slight variations in the procedure may influence the emitted sound. This variation could not be excluded completely, so measurements had to be repeated several times in order to determine the error generated through the variation in the excitation procedure.

The analysis software Praat allows to performing a reverse FFT of certain ranges of the power spectrum, which offers the opportunity to determine the decay kinetic of the amplitude of the Eigenfrequenzen or of the harmonics of an Eigenfrequenz. As the observed decay kinetic of the amplitude of an oscillation follows an exponential function, plotting the corresponding dB-values vs. time (sec) result in a linear curve. The negative slope of the resulting curve with the dimension "db/sec" is used as a parameter to define the decay kinetic of the respective oscillation.

"Resonance & Radiation spectra" (RR-spectra) were calculated from power spectra of sound samples as described elsewhere (Ref. 13). This method can also be applied to Lta-spectra. In such cases, it is of importance to define the ideal frequency-bandwidth of the Ltas in order to have a good relationship between the signal/noise ratio and the frequency-resolution of the Lta-spectrum.

Results

Most wine glasses are able to generate sounds of relatively stable intensity if they are stimulated with a finger movement on the rim of the bowl. In case the walls of the bowl are relatively thick or the glasses itself are relatively small, it might not be possible to generate a sound with such a finger movement. Another sound can be generated by hitting a wooden or metal stick against the walls of the bowl. These sounds have a typical acoustical pattern with high intensity right at the moment of the hit followed by a decline of the intensity over time.

Typical sounds generated with a hit on the bowl (hit-excitation) or a finger movement on the rim of the bowl (finger-excitation) of a wine glass are displayed in Figures 2 and 3.

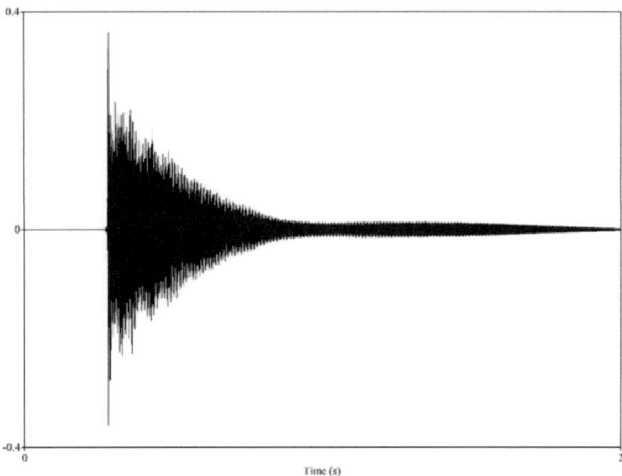

Figure 2: Oscillating wave of the sound generated by a hit with a metal stick on the bowl of the wine glass shown in Figure 1. (x-axis: time 2sec / y-axis: amplitude))

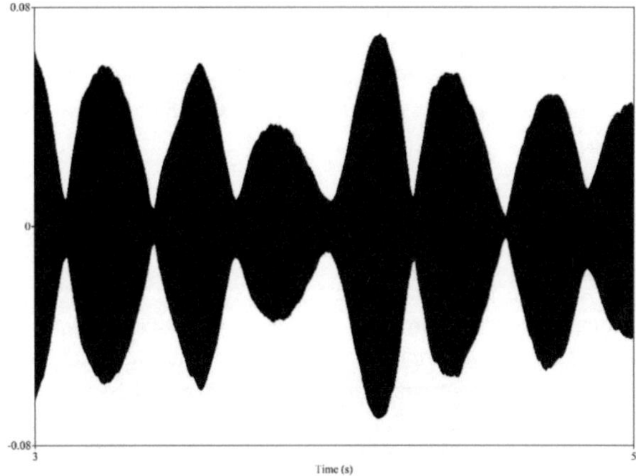

Figure 3: Oscillating wave of the sound generated by a finger movement on the rim of the bowl of the wine glass shown in Figure 1. (x-axis: time 2sec / y-axis: amplitude)

The oscillating wave displayed in Figure 3 generated by finger-excitation shows a prominent acoustic beat with a period of approximately 0,25-0,3 seconds, referring to 3,4-4Hz. The wave generated by hit-excitation displays a less prominent but detectable acoustical beat of 0,8 seconds, referring to 1,2Hz.

The corresponding frequency spectra (FFT) of the oscillating waves displayed in Figures 2 and 3 are shown in Figure 4A.

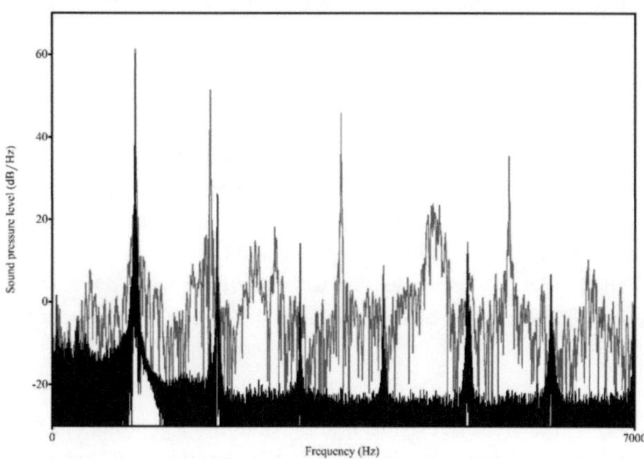

Figure 4A: Frequency spectra of oscillating waves from Figures 2 and 3 generated either by finger-excitation (black curve) or hit-excitation (red curves).

Most of the peaks in figure 4A show a structure of various closely related peaks with only minor difference in frequency (= peak-family). So the detectable peaks can be classified according to their position in the spectrum, but also according to their position within the peak-family. Table 2 summarizes and classifies the visible prominent peaks (P) of the two frequency spectra from Figure 4A, where the most prominent peak of a peak-family defines the number and the other peaks of a family are marked with a suffix.

Prominent Peaks(Hz) and corresponding dB-values			
Frequency(Hz)	dB (hit-excita.)	dB (finger-excita.)	Classification
999	52		P1B
1000,2	53,1		P1
1906,8	37,2		P2B
1912	45,4		P2
2687	19,2		P3
3484,2	40,7		P4
3489	30,1		P4B
4582	23,2		P5
5490	32,5		P6
5496,4	23,3		P6B
996,6		56,7	P1C
999,2		56,1	P1B
999,9		60,3	P1
1912		-4	P2
1995		26,5	2*P1C
2994		14	3*P1C
3994		9,1	4*P1C
4991		14,9	5*P1C
5987		6,5	6*P1C

Table 2: Frequency (Hz) and respective dB-values of the prominent peaks of the spectra displayed in figure 4A. The classification is done according to the position of the peaks in the spectrum.

It is obvious from Table 2 that among the 6 prominent Peaks of the spectrum corresponding to a hit-excitation, four of them are "double-Peaks" having two closely located dB-maxima (P1, P2, P4, P6). This confirms published findings (Ref 3-7) and is due to the various oscillation modes of a wine glass named "Eigenfrequenzen". The difference in frequency between P1 (1000,2Hz) and P1B (999Hz) of 1,2Hz is identical with the frequency of the detectable acoustic beat (see figure 1) which is caused by the overlapping effect of this two parallel oscillating waves.

With finger-excitation the most prominent peaks are in the region of the first Eigenfrequenz around 1000Hz (P1) with a weak but detectable signal of the Eigenfrequenz P2 at 1912Hz – other Eigenfrequenzen are not detectable in this case. Other types of wineglasses show also weak signals at the Eigenfrequenz P3 or very weak signals even at P4 (data not shown) during finger-excitation. Remaining peaks in the finger-excitation spectrum are harmonics of P1 (multiples of the frequency of P1C at 996,6Hz, see Table 2) which in general is a typical characteristic for a monophonic sound. Finger excitation generates three peaks (P1, P1B, P1C) at the first Eigenfrequenz with a maximum Peak-difference of 3,3Hz – this is similar to the frequency of the detectable acoustic beat of 3,4-4Hz in figure 2 caused by the overlapping of oscillating waves with similar frequencies.

The comparison of the frequency spectra of a wine glass stimulated with finger- or hit-excitation demonstrates that a finger-excitation generates more detectable peaks in the region of the first Eigenfrequenz. (see figure 5).

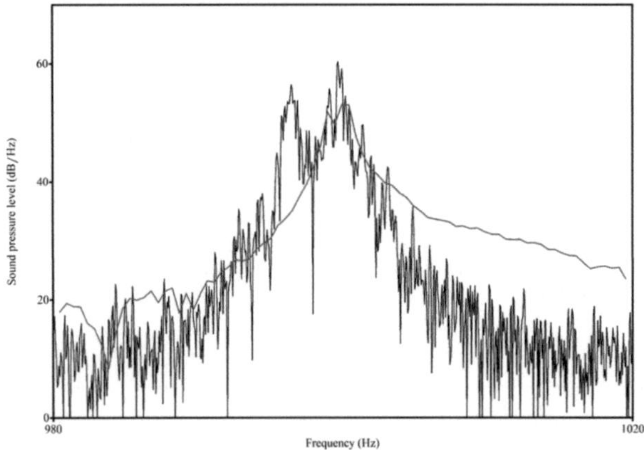

Figure 4B: Close up of the spectrum displayed in Figure 4A – range: 980-1020Hz (red-curve = hit-excitation; black-curve = finger-excitation)

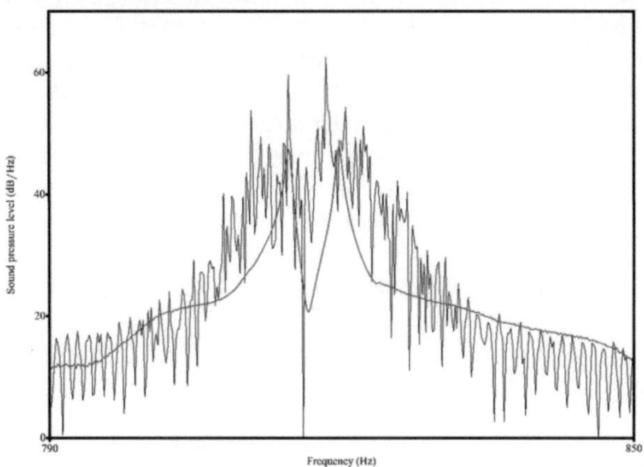

Figure 5: Spectral analysis of a wine glass (type: white wine) with focus on the area of the first Eigenfrequenz after hit-excitation (red curve) or finger excitation (black curve).

Figure 6 demonstrates that a hit-excitation can generate - beside the various Eigenfrequenzen - also a detectable signal at the harmonics of the first Eigenfrequenz. The spectrum after hit-excitation of this wine glass shows a clear peak at the 1.st harmonic at 963HZ and a weak but detectable signal at the 2.nd harmonic as well.

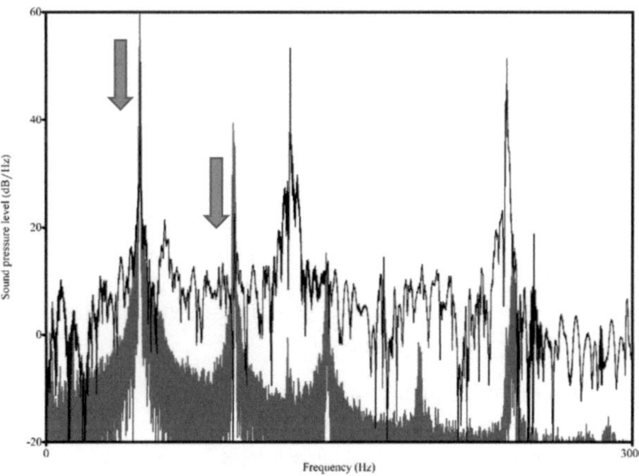

Figure 6: Spectra of the sounds emitted after hit-excitation (black curve) or during finger-excitation (red-curve) of a large "Hock"-type wine glass (see table 1). The arrows mark the peaks of the 1.st- and 2.nd harmonic of the first Eigenfrequenz.

To exclude the theoretical possibility that the respective wine glass may have Eigenfrequenzen which are identical with the frequencies of the harmonics of the first Eigenfrequenz the decay kinetics of the amplitudes of the detectable peaks in the spectrum after hit-excitation have been analyzed. As described in "Material & Methods" the decay of the dB-intensity of the various prominent Peaks in the spectrum after hit-excitation could be measured as a negative slope in "-dB/sec". This parameter describes the decay-kinetic of an oscillation. In Figure 7 the values of this decay-parameter (in "-dB/sec") are plotted against the frequency (Hz) of the prominent peaks in the spectrum of a hit-excitation (see black spectrum in figure 6).

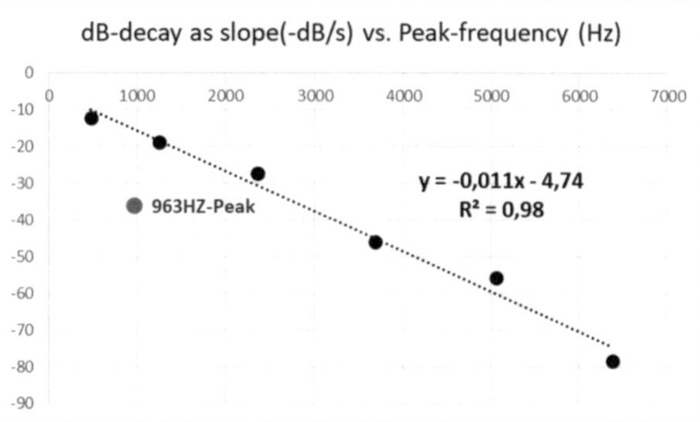

Figure 7: Calculated slopes of the linear phase of the dB-decay (in "-dB/sec) of the prominent peaks in the spectrum after hit-excitation (black spectrum in figure6) plotted against the frequency (Hz) of the peaks. The single red dot marks the dB-decay value of the peak with the frequency at the 1.st harmonic of the first Eigenfrequenz (481,5Hz). The linear regression of the decay-values of the Eigenfrequenzen (dotted line) and the respective function and R^2-value are given in the figure.

Figure 7 demonstrates the typical correlation of the frequency (Hz) of the Eigenfrequenzen and the decay kinetic of the amplitude (see linear regression function in figure 7 with R^2= 0,98). It further shows that the Peak at the frequency of the 1.st harmonic of the first Eigenfrequenz has a significant faster decay kinetic than expected for a typical Eigenfrequenz (see red dot in figure 7). This is a striking argument that the signal at the frequency of the 1.st harmonic of the first Eigenfrequenz is not an Eigenfrequenz itself, but a forced oscillation similar to the harmonics in the spectra after finger-excitation.

According to the analysis of the frequency spectra it can be stated that for most of the wine classes a hit-excitation generates a multiphonic sound where the various Eigenfrequenzen have no systematic harmonic relationship and harmonics of the first Eigenfrequenz are either not detectable or are weak in intensity. In contrast, a finger-excitation generates a monophonic sound with only minor non-harmonic signals. Therefore, wine glasses stimulated with finger-excitation can be part of a harmonic musical instrument like a glass harp. If the Eigenfrequenzen of a certain wine glass may show some harmonic relationship by generating harmonic intervals like e.g a third, a fifth, an 11th or even higher intervals, this wine glass may also be used with hit-excitation to play harmonic music.

The effect of filling liquid in the bowl on the sound of the wine glass, either stimulated with hit- or finger-excitation is different for each type of wine glass and depends on various factors. Two main effects after hit- and finger-excitation can be observed for most of the wine glass types, as a result of a changed filling status of the chalice:

1) Change of the intensity and timbre of the emitted sound and
2) Change of the frequency of the prominent peaks in the spectra.

Although the procedure of a finger-excitation of a wine glass is difficult to standardize, a high grade of repeatability of finger-excitation can be achieved through a training of the manipulating person. Reflecting this issue it will not be possible to give precis quantitative data of the effect of the filling status of the bowl on the emitted intensity of the sound, but the following qualitative descriptions may give some relevant insights:

a) By filling the bowl up to 1/4 of its entire volume with water, the measurable energy of the emitted sound can increase 3-5 fold with an increase of the dB-value for the Peaks of the first Eigenfrequenz by up to +11dB.
b) Filling the bowl with more than 1/4 of the entire volume does not result in a further increase in intensity of the emitted sound. Most wine glass types have a certain filling status somewhere at 1/3 of the entire volume, which is a "turning point" concerning the emitted sound intensity. This means that filling additional liquid in the bowl above this "turning point" results in a decline of the sound intensity.
c) As of a certain filling point of the bowl, it is no longer possible to generate a stable sound through finger-excitation.

An example for the influence of the bowl filling status on the frequency of the peaks of the first Eigenfrequenz (F1) after finger excitation is shown in Table 3.

The data in table 3 and data derived from other types of wine glasses (data not shown) demonstrate that a small amount of liquid in the chalice has only a minor effect on the frequencies of the emitted prominent signals of the first Eigenfrequenz generated by finger-excitation – which confirms previously published data (Ref. 7; 8). Increasing the filling status leads to significant reductions of the frequencies of such peaks. For most wine glasses, a filling status of 50% or higher reduces the frequency of the prominent peaks of the first Eigenfrequenz by 10% or even more (data not shown) before a further increase of the filling status stops a stable sound production through finger-excitation. The frequency of the harmonics of the first Eigenfrequenz (F1) change accordingly, as those are integer multiples of the frequency of F1.

Frequency (Hz) of prominent Peaks / finger-excitation			
Peaks	**Empty**	**50ml/ 10%**	**110ml/22%**
F1A	622,2	621,8	617,2
F1B	625,5	625	618,7
F1C	632,2	628,2	622
F1D	636	634,2	628

Table 3: Frequency (Hz) of the four in the spectrum separated and prominent Peaks of the first Eigenfrequenz (F1) after finger-excitation of an empty and partly filled Bordeaux-type wine glass. 50ml and 110ml reflect 10% and 22% of the entire volume of the chalice.

The influence of the filling status of the bowl on the frequencies of the Eigenfrequenzen of a Bordeaux wine glass stimulated through a hit-excitation is summarized in table 4.

Frequency-shift of Eigenfrequenzen due to filling status					
Filling(ml)	**F1B(Hz)**	**F2A(Hz)**	**F3A(Hz)**	**F4A(Hz)**	**F5A(Hz)**
0	634,2	1696	2637,2	3047,1	4023,3
50	633,3	1695,9	2558	3046,9	4022,9
110	629,2	1694	2597,3	3044,8	4019,8
200	604,9	1663,2	not detected	2933,6	3691,7

Table 4: List of frequencies of the Eigenfrequenzen (F1-F5) of a wine glass (Bordeaux type) stimulated with a hit-excitation. The bowl was either empty or filled with 50ml, 110ml or 200ml of water prior to the hit.

The frequencies of F1, F2, F4 and F5 show very similar effects: 50ml in the chalice has only a minor effect on the frequency, but increasing volumes result in stronger reductions of the frequency. The Eigenfrequenz F3 shows a different pattern and in case of 200ml this signal is not detectable in the spectrum after hit-excitation.

Filling	**dB-value**
Empty	55,6
50ml	62,2
110ml	61,2
200ml	62,3

Table 5: dB-values of the emitted sound (measurement period 5sec) after hit-excitation of a Bordeaux wine glass filled with 0, 50, 110 and 200ml of water. The error for the measurement of the intensity is $\pm1,5$dB.

An example for the influence of the filling status of the bowl with water on the intensity of the emitted sound after hit-excitation of a wine glass is demonstrated in table 5.

For most types of wine glasses, a change of the filling status of the bowl has some influence on the timbre of the emitted sound during finger excitation. Table 6A demonstrates that the

"Center of Gravity" (which is a parameter describing the timbre of a sound) is shifted towards a higher frequency with increasing filling status. Although the frequency of the dominating Eigenfrequenz of the emitted sound during finger-excitation is shifted towards lower frequency with an increasing filling status of the bowl (see table 3) the timbre develops in the opposite direction which means that the overall emitted energy is shifted to oscillations of higher frequency.

Filling (ml)	CG-2.0(Hz)
0	671
50	711
100	683
200	915
300	1262

Table 6A: Values for "Center of Gravity 2.0" (CG-2.0) in Hz of the emitted sound (measurement period 5sec) after finger-excitation of a Bordeaux wine glass (same as in table 5) filled with 0, 50, 100, 200 and 300ml of water reflecting a bowl filling status of 0%, 10%, 22%, 40% and 60%. The error for the calculation of "CG-2.0" is $\pm$ 25Hz.

Table 6B demonstrates that the "Center of Gravity" of the emitted sound by a Bordeaux type glass after hit-excitation shows a shift towards higher frequency with an increasing filling status of the bowl (similar to the effect observed during finger-excitation) although the signals derived from the dominant Eigenfrequenzen up to 4100Hz have lower frequencies with an increasing filling status of the bowl (see table4).

Filling(ml)	CG-2.0 (Hz)
0	5421
50	6258
110	6000
200	8092

Table 6B: Values for "Center of Gravity 2.0" (CG-2.0) in Hz of the emitted sound (measurement period 5sec) after hit-excitation of a Bordeaux wine glass (same as in table 5) filled with 0, 50, 110 and 200ml of water which reflect a filling status of 0%, 10%, 22% and 40%. The error for the calculation of "CG-2.0" is $\pm$ 100Hz for a filling of 0-50ml and $\pm$250Hz for 110-200ml.

An increasing filling status shifts the Central Gravity of the emitted sound towards higher frequencies for both types of glass-excitation, although the peaks of the Eigenfrequenzen are shifted in the opposite direction towards a lower frequency. This means that an increase in the amount of liquid in the bowl has the expected damping-effect on the frequency of the oscillation of the Eigenfrequenzen, but also shifts the emitted acoustic energy towards the Eigenfrequenzen and harmonics of higher frequency. This is demonstrated well by comparing the calculated "Resonance & Radiation spectra" (Ref. 13) of the sounds generated with finger excitation of a wine glass filled with 0ml (empty) or 300ml water (see figure 8).

It is obvious from figure 8 that a filling status of 60% (30ml) vs. an empty bowl generates besides the general shift of the resonances towards lower frequency a very strong resonance in the region of 4000Hz and a further increase of the resonance at around 9000Hz. Those massively increased resonances after filling 300ml into the bowl could be responsible for the shift of the Central Gravity of the emitted sound towards a higher frequency during finger excitation.

Although the method to calculate a "Resonance& Radiation spectrum" (RR-spectrum) is described for monophonic sounds with one basic frequency and related harmonics, this method of calculation can also be applied to Lta-spectra of multiphonic sounds (see Material & Methods). For this calculation, the defined bandwidth of the Ltas is of critical importance, as it influences the resolution of the corresponding "Resonance & Radiation spectrum". In order to obtain a RR-spectrum for hit-excitation of a wine glass with a similar resolution as for a RR-spectrum of a finger-excitation, the bandwidth of the Ltas should be similar to the frequency (Hz) of the dominating Eigenfrequenz (visible in the power spectrum of a finger excitation). This allows a good comparison of the two RR-spectra, which is demonstrated in figure 9 for an empty Bordeaux type wine glass.

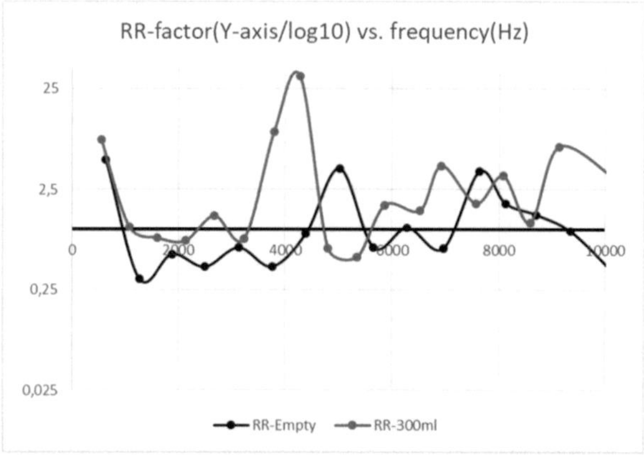

Figure 8: Plot of the calculated "Resonance & Radiation factors" vs. frequency (see Material & Methods) for a Bordeaux type wine glass filled with 0ml (Empty-black curve) or 300ml water (red curve) and stimulated with a hit. The RR-factor has no dimension and the y-axis has a logarithmic scale as e.g. a value of 0.25 for the RR-factor has the same but opposite meaning as a value of 4.0 (see Material and Methods)

Although the grade of frequency-resolution of the RR-spectra (hit- and finger excitation) displayed in figure 9 is relatively low, still areas of high resonance of the respective wine glass are detectable as peaks in the spectra. Up to 10.000Hz seven prominent resonance-areas can be identified with a good match between the spectra generated with hit- or finger-excitation (considering the low frequency-resolution) except for a missing signal in the range of 6000-6500Hz in the "hit-spectrum" vs. the "finger-spectrum".

In some cases, the manipulating person executing the finger-excitation of a wine glass can generate a sound with a frequency-spectrum showing characteristic features of a typical hit-excitation as well as of a typical finger-excitation. Such a spectrum is displayed in figure 10. It is obvious that is this case harmonics of the dominating peak are detectable as well as various other peaks which cannot be classified as harmonics of the prominent peak.

In figure 11 the classification of the detectable maxima (peaks) from the spectrum in figure 10 is given. The dominating peaks are those of the Eigenfrequenz F2 (622-636Hz) and their related harmonics - which is a typical feature of a sound generated through finger excitation. Beside the dominating signal of F2 the Eigenfrequenz F1 (180Hz) and its 1.st harmonic (360Hz) is clearly visible as well as signals of the Eigenfrequenzen F3 (1695Hz), F4 (3050Hz) and F5 (4028Hz) – this is a typical feature of a sound generated through hit-excitation. The remaining weak but detectable signals in the range up to 4500Hz can be classified as "Complex tones" as their frequencies are derivatives of the frequencies of F1 and F2 and therefore show the typical characteristics of complex tones described for multiphonic sounds generated with woodwind instruments (Ref.10).

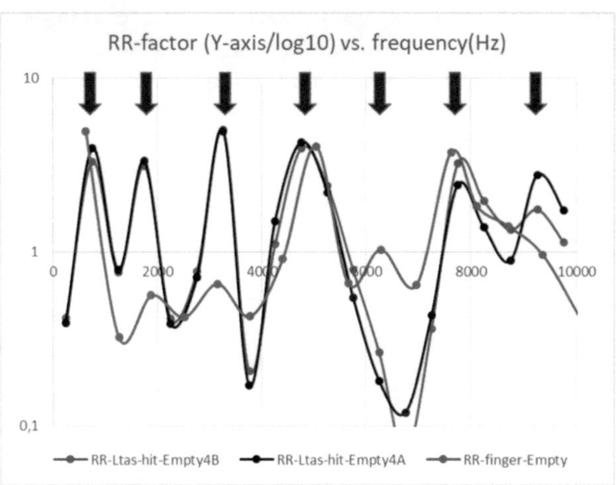

Figure 9: Plot of the calculated "Resonance & Radiation factors" (RR-factors) for an empty Bordeaux type wine glass vs. frequency (Hz). The red curve (resolution: 310Hz) shows the RR-factors for a finger-excitation from figure 8. The black and blue curves (resolution: 250Hz) show the RR-factors calculated from Lta-spectra (bandwidth of Ltas: 500Hz) of two sounds (named: 4A & 4B) generated by hit-excitation (see Material & Methods). The blue arrows mark frequency-areas with visible resonance signals.

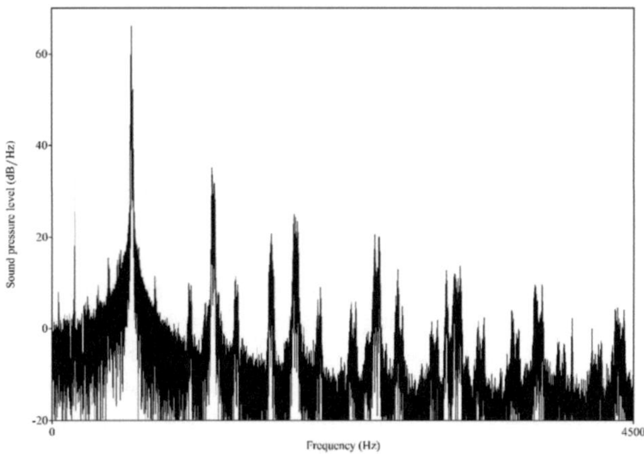

Figure 10: Spectrum derived from a finger-excitation of a wine glass (Bordeaux type). The manipulating person adjusted the finger-pressure on the rim of the bowl as well as the speed of finger-movement in such a way that the resulting stable sound contains features of a typical sound after finger-excitation, but also features of a typical sound generated through a hit-excitation.

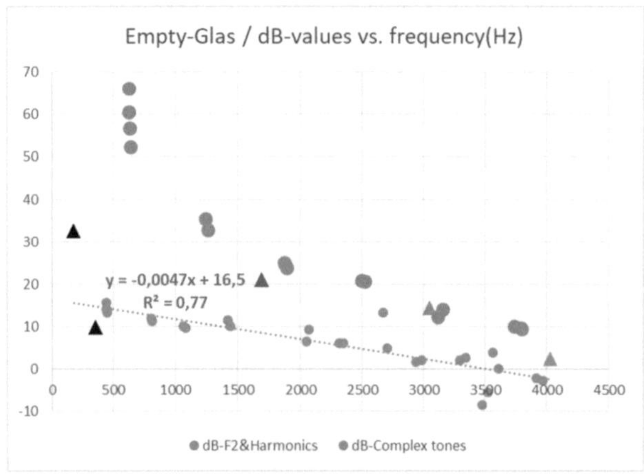

Figure 11: dB-values of the detectable maxima from the spectrum in figure 10 are plotted against their frequency (Hz). The prominent signals of the Eigenfrequenz "F2" and its related harmonics are marked with blue dots. The black triangles mark the Eigenfrequenz "F1" and its 1.st harmonic. The red, brown and green triangles mark the Eigenfrequenzen F3, F4 and F5. The complex tones (derivatives of F1 and F2) are marked as brown dots – the respective linear regression function of the complex tones and the value for R^2 are given in the graph.

Although it has been found that for most wine glasses the first Eigenfrequenz (including its related harmonics) is the oscillation-mode which is dominantly stimulated by finger-excitation, it is demonstrated in figures 10 and 11 that also the oscillation-mode of the 2.nd Eigenfrequenz can be the signal dominantly stimulated through a finger-excitation.

Discussion

It has been demonstrated that "wine glasses" can behave like monophonic instruments with an adequate finger-excitation (see figures 3; 4) but generate multiphonic sounds due to a hit-excitation (see figures 2; 4). The power-spectrum of a sound generated with a hit-excitation displays the various "Eigenfrequenzen" of the wine glass with its characteristic double-peaks as described elsewhere (Ref. 5-8). During finger-excitation, one of the Eigenfrequenzen (in most cases the "Eigenfrequenz" with the lowest frequency) builds the base-tone with several detectable harmonics which are frequency-multiples of the base-tone. The other "Eigenfrequenzen" are either very weak or are not detectable during finger-excitation. This could be interpreted as 1) a massive damping of the other Eigenfrequenzen by the finger which slides on the rim of the bowl and 2) a strong excitation of the specific Eigenfrequenz which builds the base-tone of the resulting monophonic sound.

The data presented in figure 7 indicate that the type of oscillation of the glass-wall is different between the Eigenfrequenzen and the harmonics of a base-tone, as the decay characteristic of the 1.st harmonic differ significantly from those of the Eigenfrequenzen. This aspect should be investigated and analyzed in further research.

It is common practice of musicians playing glass-harps to fill some water into the bowl of the wine glass in order to tune the prominent Eigenfrequenz of the glass (being the base-tone of the generated monophonic sound) to the desired frequency. As filling of liquid into the bowl will not only lower the frequency of the base-tone and the respective harmonics as described previously (Ref. 5-7) but also changes the intensity as well as the timbre of the generated sound significantly (see "Results", - tables 6A; 6B and figure 8) musicians must consider these effects as well when tuning wine glasses by filling the bowl with water. In general, it can be concluded that additional water acts as an additional mass and therefore has a damping effect which is stronger at oscillations of lower frequency and gets weaker with increasing frequency of the oscillations.

It is worth to note that beside the type of manipulation by the musician ("playing" the glass harp through finger-excitation) the specific resonance-characteristics of a certain wine-glass with its various Eigenfrequenzen also determines the monophonic sound generated by the musician (see figure 9).

Under certain circumstances of a finger-excitation, the resulting sound shows features of a monophonic sound (base-tone and related harmonics) and features of a multiphonic sound (frequencies with no harmonic relationships plus complex tones) in parallel (see figures 10; 11). This results in a power spectrum which is highly similar to power spectra of multiphonic sounds generated by wood wind instruments (Ref. 10).

If we classify a wine glass as a "musical instrument" being able to produce monophonic sounds through finger-excitation and multiphonic sounds through hit-excitation (see figures 2-4) we may see similarities to wood wind instruments which are also known as monophonic instruments being able to produce controlled multiphonic sounds (Ref. 10, 11). As the "construction" of a wine glass is by far less complex than a wood wind instrument it might be a good object to investigate the principle mechanism of the production of complex tones within multiphonic sounds generated by "tube-like" instruments as wood winds.

References

1) website: https://christnersprimesteakandlobster.com/which-glass-for-which-wine-wine-glass-guide/; download: 6/2022

2) L.Mäder, L.Moheit, L.Kastenhuber and S.Marburg; Zur Akustik von Weingläsern; Akustik Journal 03/18, pp 7-14; 2018

3) H.Schlichting; Glockenklang im Weinglas; Naturwissenschaften im Unterricht –Physik 39/10; Vol. 20; 1991

4) H.Schlichting; Es tönen die Gläser; Physik in unserer Zeit; Vol. 26/3; pp 138-139; 1995

5) G.Denninger; Das Ohr trinkt mit; Physik in unserer Zeit; Physik in unserer Zeit; Vol. 44; pp 142-146; 2013; DOI: 10.1002/piuz.201301327

6) W.Poomarin and I.Jacobs; Wine glass oscillations; website: https://kvis.ac.th/userfiles/files/Wine%20glass%20oscillations.pdf; download 2/2023.

7) S.Yang; Wine glass acoustics; website: https://studylib.net/doc/18265335/wine-glass-acoustics-how-does-the-frequency-vary-with-the..; download: 2/2023

8) Cosmol software-App, website: https://apps.vib.ed.tum.de:2037/comsol-software-license-agreement; date of usage: 2/2023

9) A.French; In Vino veritas – A study of wine glass acoustics; Am. J. Physics; Vol. 51 No.5; 1983

10) A.Rehm; Analysis of spectral parameters of mulitphonic sounds of string- and wind-instruments and introduction of a mathematical model to simulate the frequencies and intensities of the harmonics and the complex tones of multiphonics generated with wood winds; ISBN: 97833466652799; 2022; Deutsche Nationalbibliothek; http://dnp.d-nb.de

11) J.Backus, Multiphonic tones in the woodwind instruments. The Journal of the Acoustical Society of America, 63(2), pp 591–599; 1978.

12) References and details about "Praat" software see website: https://fon.hum.uva.nl/praat/ ; status 1/2022.

13) A.Rehm; Presentation of the "Resonance & Radiation" spectrum (RR-spectrum) as a parameter to describe the sound characteristics of musicians playing monophonic instruments; ISBN: 9783346591654; 2/2022; Deutsche Nationalbibliothek; http://dnp.d-nb.de